D0082868

HIGH PERFORMANCE CONCRETES AND APPLICATIONS